AF577158

ECCENTRIC MACHINES

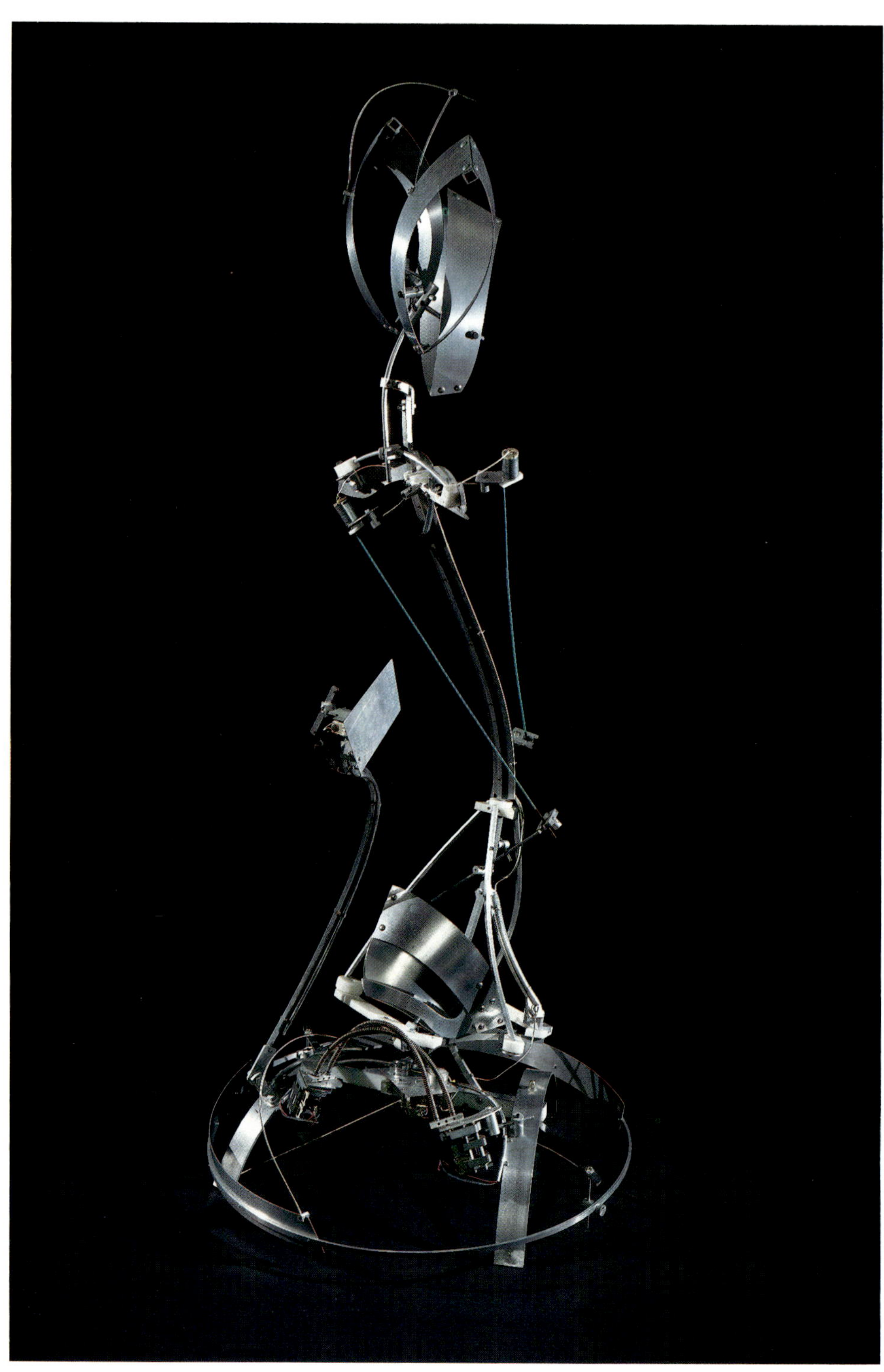

16 Gary Justis, *Chromosome,* 98 x 40 x 40"

ECCENTRIC **MACHINES**

JOHN MICHAEL KOHLER ARTS CENTER
SHEBOYGAN, WISCONSIN

JOHN MICHAEL KOHLER ARTS CENTER STAFF

Ruth DeYoung Kohler, director
Joanne Cubbs, curator of exhibitions
Jean Lorrigan Puls, associate curator of performing arts and education
Emil L. Donoval, registrar
Mary Jo Ballschmider, manager-administrative services
Robert Teske, associate curator of exhibitions
Sheila Webb, associate curator
Richard Zauft, Arts/Industry coordinator
Eric Dean, Resource Center coordinator
Juli Leet, publicity administrator
Sally Bohenstengel, accountant
Roger Krueger, physical plant supervisor
Sara Bong, assistant to the director
Dorothy Hellauer, word processor
Jane Andrews, secretary/receptionist I
Eric Johnson, technical coordinator
Ron Krepsky, technician
Deborah Sandven, Signature Gallery coordinator
Nancy Dummer, administrative assistant
Denise Lau-Kuehlmann, Preschool administrator/head instructor
Nancy Zeller, assistant Preschool instructor
Darla Faken, bookkeeping assistant, secretary/receptionist II
Trudy Gruber, secretary/receptionist II
Dang Yang, security guard, senior aides program
Kenneth Hauch, security guard, senior aides program

LIBRARY OF CONGRESS CATALOGING-IN-PUBLICATION DATA
Eccentric Machines

Catalog of the exhibition organized by the John Michael Kohler Arts Center, held June 14 - August 23, 1987.

1. Mechanized art — Exhibitions. 2. Art, Modern — 20th century — Exhibitions. I. John Michael Kohler Arts Center.

N8222.M27E3 1987 709'.04 87-31100 ISBN 0-932718-22-1

Published on the occasion of an exhibition organized by and at the John Michael Kohler Arts Center, 608 New York Avenue, Sheboygan, Wisconsin, 14 June - 23 August 1987. Supported in part by grants from the National Endowment for the Arts and the Wisconsin Arts Board.

A portion of the Arts Center's general operating funds for this fiscal year has been made available through a grant from the Institute of Museum Services, a federal agency that offers operating and program support to the nation's arts centers and museums.

Printed in the United States of America.

Note: Within the text, references to catalogue numbers appear in brackets.
Cover: Gary Justis, *Chromosome* detail (16).

CONTENTS

Figure 1. Installation view of ECCENTRIC MACHINES exhibition.

PREFACE & ACKNOWLEDGMENTS

During the summer of 1987, the John Michael Kohler Arts Center organized and presented ECCENTRIC MACHINES, a major exhibition exploring the use of mechanization in contemporary American art. It featured twenty-six large-scale sculptures by twenty artists from seven states, including Illinois, Arizona, New York, Georgia, Washington, California, and Michigan. The exhibition and this publication celebrate a unique genre of art making today and offer a fascinating new chapter in the long history of the relationship between art and the machine.

The gala opening of ECCENTRIC MACHINES featured a variety of programming. Members of Kanopy Dance Theatre offered an entrancing improvisational performance in the Main Gallery in response to the mechanized art in the exhibition. A workshop exploring synthesized sounds was conducted by local artist Kevin Zimmermann for adults and children. Interpretations of two Ray Bradbury stories about the eerie effects of mechanization on our lives were dramatized by area students. A gallery talk was presented by Curator of Exhibitions Joanne Cubbs, and the films *Ballet Mechanique* (Fernand Leger, 1924) and *Ballet Robotique* (Bob Rogers, 1983) were shown continuously throughout the afternoon.

During the course of ECCENTRIC MACHINES, a number of related projects were developed, the most complex of which was *Mechanical Marvels.* This huge mechanized installation was created on the Arts Center's east lawn by Georgia artist Bryant Baker, Milwaukee sculptor Chuck Toman, and Sheboygan artist Susan Olsen Pardo in collaboration with thirty adults and 112 children from the community. Baker headed the project during a five-week residency at JMKAC. Working literally day and night, he and the other participants used an enormous variety of found and donated materials — from milk crates to car parts — and electrical components to create a large "house" that whizzed and whirred and clanked plus a variety of life-size and over-life-size creatures. Among them was a "dog" attempting to attack a "cat." An "ostrich" flapped its wings and bobbed about while its back sparkled with lights. Another long-legged creature had a large magnet spinning in its transparent torso which made its iron-filing "hair" ruffle and stand on end. After its completion, the *Mechanical Marvels* site work remained on display throughout the summer.

Other programming included Gadgets and Gizmos, during which individuals from the area shared their treasured mechanical collections with the public. The Great Cardboard Boat Regatta featured mechanized corrugated boats that were both artistically fascinating and surprisingly fleet. Second- through fifth-grade students concentrated on the concepts of mechanized art — using the ECCENTRIC MACHINES exhibition as their initial source of study and inspiration — to create performance pieces as well as a host of six- to seven-foot robots. Even the Seventeenth Annual Outdoor Arts Festival focused on the mechanized theme in a workshop and demonstrations.

The exhibition ECCENTRIC MACHINES, its accompanying publications, and associated programming were made possible through the contributions of many individuals and organizations. We gratefully acknowledge the National Endowment for the Arts, the Wisconsin Arts Board, and the David Bermant Foundation: Color, Light, Motion as well as a number of Sheboygan area corporations and foundations for providing financial support for the project.

We express our deep appreciation to the twenty contemporary American artists who participated in the exhibition. In addition, we thank the following galleries and individuals who loaned works to the exhibition or assisted with arrangements for them:

Marianne Deson Gallery, Chicago, Illinois
Martin Edelston, New York, New York
Gracie Mansion Gallery, New York, New York
Paul Klein Gallery, Chicago, Illinois
Sander Gallery, New York, New York
John Weber Gallery, New York, New York

We are also indebted to the many individuals and organizations in Wisconsin who have contributed their talents, time, materials, and equipment to the exhibition and its related programming:

Akright Auto Parts, Sheboygan Falls
Allen-Bradley Co., Milwaukee
AmeriGas, Industrial Gases Division, Sheboygan
AmeriGas Welders World, Sheboygan
Bemis Manufacturing Company, Sheboygan Falls
Alan Bengston, Sheboygan
Bike 'n Ski Warehouse, Sheboygan
Bitter-Neumann & Co., Howards Grove
Scott Bohenstengel, Sheboygan
Judy Born, Sheboygan
Dick Brantmeier Ford, Inc., Sheboygan
City Bakery, Sheboygan
Cleveland Auto Sales and Salvage, Inc., Cleveland
Shelly Dekarske, Sheboygan
Marvin DeSwarte, Two Rivers
Diamond Vogel Paint Center, Sheboygan
Marie Felzo, Sheboygan
The Foley Co., Manitowoc
Formrite Tube Co., Inc., Two Rivers
Garton Metal Finishing Company, Sheboygan
John Green, Kohler
Jeffrey Hildebrand, Kohler
Gus Holman Company, Sheboygan
R. P. Honold Co., Inc., Sheboygan
Imperial Clevite, Manitowoc
Ann Marie Jacobson, Sheboygan
Julie Johnson, Sheboygan
Tim Johnson, Sheboygan
Johnston Bakery Inc., Sheboygan
Kaltenbrun Bros. Roofing Co. Inc., Sheboygan
Kohler Co., Kohler
Kohler General, Inc., Sheboygan Falls
Lakeshore Technical Institute Robotic Lab, Cleveland
Casper Meinhardt, Sheboygan
Chuck Meives, Sheboygan
Muchin Steel Supply Co., Inc., Manitowoc
Keith Nilson, Sheboygan
Paragon Electric Co., Inc., Two Rivers
Greg Pardo, Sheboygan
Plastics Engineering Company, Sheboygan
Pomp's Tire Service, Inc., Sheboygan
Brian Reindl, Manitowoc
Dave Rhineg, Sheboygan
Richardson Lumber Company, Sheboygan and Sheboygan Falls
Janet and Carl Ross, Sheboygan
Susan Roy, Sheboygan
Jos. Schmitt & Sons Construction Co., Inc., Sheboygan
Sheboygan Area Forensics Group, Sheboygan
Sheboygan Falls Department of Public Works, Sheboygan Falls
Sheboygan Glass Co., Sheboygan
Sheboygan Paint Company, Sheboygan
Sheboygan Scrap Metals, Sheboygan
Sherwin-Williams Co., Sheboygan
Martha Spiller, Oostburg
Spiller Spring Company, Sheboygan
Brad Suhm, Sheboygan
Triangle Used Auto Parts, Sheboygan
Universal Lithographers, Inc., Sheboygan
The Vollrath Company, Sheboygan
Bob Werner Chevrolet-Cadillac, Inc., Sheboygan
Tyrone Wesley, Sheboygan
Julie Wickland, Sheboygan
E.C.H. Will, Inc./Pemco Co., Sheboygan
Wolf Cycle Center, Sheboygan
Zona Ziegler, Sheboygan
Kevin Zimmermann, Cedar Grove

Finally, we thank the Arts Center staff for their expertise and dedication. ECCENTRIC MACHINES was curated by Joanne Cubbs who must be recognized for developing an exhibition with both critical significance and strong public appeal. Larry Donoval ably arranged for the shipping, insurance, and installation of exhibition objects, tasks complicated by the demanding mechanical nature of the works. Roger Krueger, Eric Johnson, and Ron Krepsky also played an important role in the complex construction and installation of the exhibition. Jean Lorrigan Puls coordinated the *Mechanical Marvels* project, Arts Day Camp, and the opening day celebration. Mary Jo Ballschmider administrated the Great Cardboard Boat Regatta, and Sara Bong coordinated the Gadgets and Gizmos event. Juli Leet orchestrated media coverage, and Eric Dean supervised photographic documentation. Sheila Webb coordinated the production of this publication, while Richard Zauft stepped in to execute its design. Sally Bohenstengel carefully administered the project budget. Nancy Dummer proofed all copy, while Deborah Sandven helped with organizational details. Dorothy Hellauer and Jane Andrews devoted countless hours to typing correspondence, publication copy, labels, and educational materials for the project. Ken Hauch and Dang Yang enthusiastically aided thousands of visitors with the operation of the mechanized works of art.

The teamwork and active involvement of artists, volunteers, contributors, Board, and staff have been critical ingredients in realizing the ECCENTRIC MACHINES exhibition and programming. Our boundless appreciation to all of you.

Ruth DeYoung Kohler

11 Rodney Alan Greenblat, *I Saw the Channel 5 in Gold*, 66 x 66 x 7."

ECCENTRIC MACHINES

The machine allows me, above anything, to reach poetry.

Jean Tinguely

Since the late nineteenth century, nothing has affected human existence more dramatically than the machine. By the 1920s, mechanization had begun to pervade even the most intimate areas of American life. Household appliances and the automobile came into widespread popular use. Across the country, cities continued to grow, and everywhere there appeared the giant metal apparitions of a truly industrial civilization. A proliferation of factories, grain elevators, bridges, and skyscrapers created a new native landscape, and with it came a new perception of reality, one inextricably tied to the idea of the machine.

Not surprisingly, the machine emerged as a strong leitmotif in the art of the early twentieth century. Beyond the representations of machines in paintings, drawings, and sculpture, a number of artists produced works that incorporated real mechanisms. In 1920, the Russian Constructivist Naum Gabo created the mechanical sculpture *Virtual Kinetic Volume.* During that same year, Gabo and his brother Antoine Pevsner issued the following proclamation in the *Realist Manifesto:* "We reject the thousand-year-old illusion in art that a static rhythm is the only element of the plastic arts. We affirm in these arts a new element, kinetic rhythm." In 1922, another Russian, Alexander Archipenko, invented an electrically driven machine for creating moving pictures called an *Archipentura.* Around the same time, Marcel Duchamp began a series of rotating apparatuses which produced unusual optic effects; Laszlo Moholy-Nagy created a motor-driven sculpture which engaged viewers in a spectacle of moving light; and Man Ray mounted the photograph of an eye to a metronome and called it *Object of Destruction.* Finally, in the 1930s, Alexander Calder produced his now well-known mobiles, propelled first by small electric motors or tiny handles and later by currents of air.

During the first decades of the century, the machine came to illustrate ideas central to the philosophies of several movements in art, including Dadaism, Russian Constructivism, Italian Futurism, Vorticism, Purism, Precisionism, and Surrealism. The meaning ascribed to this symbol of technology varied greatly. The Futurists, for example, aggressively worshiped the machine as a new modern god. They exalted speed and technology as a source of "new beauty." With similar messianic fervor, the Contructivists viewed the machine as a harbinger of social and technological utopia. In contrast, the Dadaists used mechanical forms to raise disconcerting questions about the interrelationships between humanity and the machine. Their anthropomorphic machine portraits promulgated a vision of humankind as beings without will or self-determination. Their ironic transformation of found machines into nonfunctional art objects pointed to the irrationality underlying existence in an increasingly mechanized and materialistic world.

In the period between the two world wars, often referred to as the "age of mechanization," the machine had seemingly reached its height as an icon of twentieth century art. Yet, during the decades of the 1950s and 1960s, a number of artists once again began to explore the expressive possibilities of motion. The term "kinetic art" was widely adopted to describe art that incorporated real movement through mechanical or other means and frequently in combination with auditory and light effects. Many artists applied the latest scientific methods and technologies to their work and formed creative liaisons with scientists and engineers.

Among the numerous groups aspiring to an integration of art and science were the Groupe de Recherche d'Art Visuel in France, Experiments in Art and Technology, Inc. (EAT) in the United States, and in London, the Centre for Advanced Study of Science in Art.

Through their experiments, kinetic artists continued to expand the traditional definition of art. Their broad interdisciplinary approach prompted new adventures into time/space relationships, light, color, sound, cinema, and performance and installation art. Their curious machines and fantastic spectacles used the drama of movement to involve viewers more directly with their artistic ideas. Pol Bury's three-dimensional constructions mystified spectators with their hidden mechanisms and unpredictable, nearly imperceptible motion. In a similar way, the artist Takis used the invisible forces of gravity and magnetism to create unexpected movements in his steel and iron sculptures. Nicholas Schöffer was one of the first to use electronics and cybernetics as a means of integrating movement with light projections and music. With a Duchampian sense of irony and humor, Jean Tinguely created machine contraptions which satirized both art and technology. His whimsical coin-operated devices for making pictures, his mechanized assemblages of industrial debris, and his infamous self-destructing machine sculptures embodied a sense of surprise and chance happening. Like many others, Tinguely's purpose was not to create "art objects" but rather "art events."

Today in a postindustrial age of computers, lasers, microcircuitry, and compact discs, the machine continues to incite artists' imaginations. ECCENTRIC MACHINES presents a wide range of contemporary mechanized art, from simple hand-cranked devices to electronic sculpture and state-of-the-art robotics. Underlying this idiosyncratic assembly of moving parts, blinking lights, and obscure sounds is a strong feeling for the mythology of meaning which still surrounds the machine. Also evident is a powerful sense of irony as each artist subverts the utility, logic, and uniformity associated with the machine to create his or her own eccentric vision.

A number of works focus on the physical qualities of movement. Dean Langworthy's rudimentary viewer-operated devices offer a poignant lesson in mechanical cause and effect. In his piece *Allegory Engine Series No. 5* (18), the spectator sits on a stool and turns a crank in order to propel himself/herself upward along two heavy steel rails propped at an angle against the wall. Because the participant must crank wildly to move even a short distance, the artist regards his ponderous mechanism as an allegory for "the absurdity and futility of effort spent in seemingly noble pursuit." Langworthy states: "I am trying to present the storytelling possibilities of movement, mass, power, and the performance of work. There is no illusion in my pieces; the properties they have are real; the forms and actions are not abstracted; they simulate nothing; they are the machines they seem to be." The simplicity and literalness of Langworthy's "primary machines," as well as his attempts to heighten the physical and psychic perceptions of the viewer, recall the formal strategies and phenomenological concerns of Minimal Art.

Lewis Alquist's motorized assemblages also make the spectator acutely aware of certain physical phenomena. In his *States of Matter Undergoing Habitual Rotation* (4), Alquist uses movement to create the illusion of matter's transformation into solids, liquids, and gases. The piece features two identical dwarfed bed frames which spin counterclockwise at different velocities. The slower bed holds a glass liner filled with a white opaque liquid that appears to form a solid shape while under rotation. The second bed, rotated at a speed high enough to make its form indistinguishable, takes on the appearance of a gas. In addition to enacting such pseudoscientific demonstrations, Alquist's machine sculptures attempt to capture a feeling for the comedy and terror underlying existence. The violently spinning beds in *States of Matter* are strangely unnerving.

In contrast to Alquist's frequent use of an aggressive spinning motion, Joe Cavalier animates his abstract sculptures with a slow, undulating rhythm. This quiet, even, meditative movement causes a subtle shifting in the relationship of forms. *Moon/Lightning* (8) is a triangulated web of fine rods which defies its own rigid geometry with a gentle waving motion. This sensuous and organic movement serves as an intriguing foil to the rational clarity of Cavalier's engineered constructions and, according to the artist, "turns the dynamics of the machine inside out."

A significant number of artists use the machine as a metaphor for otherwise intangible, abstract subjects — like the human psyche, mystico-religious phenomena, or the principles of biological life. In probing the metaphysical possibilities of the machine, these artists have discovered a new world, one which exists somewhere between the realms of art and technology, science and magic. Often underlying their works is a shared belief that the physical energy of the machine can function as a positivist equivalent to more invisible forces. Their common association of electricity with human thought, divine spirit, and other transrational powers finds its precedence in Henry Adams's famous ontological comparison of the Virgin and the dynamo made upon his visit to Paris and the Great Exposition's gallery of machines in 1900.

Dennis Oppenheim, who has been making mechanized art since 1979, invents machine metaphors for the workings of the human mind. The mechanics of thought and, in particular, the creative process have intrigued him since his involvement with Conceptual Art in the early 1970s. In *The Day Before Starry Night (For Vincent Van Gogh)* (20), Oppenheim evokes the emotional trauma of artistic creation through a trembling shower of curled metal strips and tiny white lights which flow down from beneath a giant outstretched hand. Intended to reenact the anxious moment just before Van Gogh's execution of his well-known masterpiece *Starry Night,* Oppenheim's work possesses a feeling of tense, contained energy. In many of the larger mechanized environments or "factories" which Oppenheim has also been building since the early 1980s, his machines are not contained; in fact, they have gone completely mad. A combination of technology and theatricality, these huge pseudo-industrial structures spin, shake, smoke, spark, shoot fireworks, and self-destruct.

Alice Aycock also captures a feeling of violent, chaotic power in a number of spinning blade sculptures that she began making in 1982. One work from the series titled *Fata Morgana (Of Things Seen in the Sky)* (5) employs giant spinning blades, a streak of orange neon light, and white sparks from a grinding wheel to create a sense of invisible cosmic forces. Aycock, who often links modern technology with mysticism and magic, drew inspiration for *Fata Morgana* from medieval accounts of Morgan Le Fay, the sinister sorceress and half sister of King Arthur. The title of the work refers more specifically to the strange floating mirages often conjured by the enchantress. The mirror suspended just above the artist's motorized sculpture is a clever reincarnation of Morgan Le Fay's hovering apparitions. Such spectres are frequent protagonists in Aycock's mechanized works.

The electrokinetic aluminum sculptures of Gary Justis possess a strong mythic quality. These robotlike figures serve as surrogate performers in dramas of the human psyche. Their heroic scale and enchanting formal beauty increase the feeling that they are actors of mythological proportion. In *Controlled Stamen* (17), a giant mechanical flower lights up and begins to shake, quiver, and vibrate in strange erotic excitement. Beside it a small metronome (the infamous "biological clock") stoically marks the time completely unaffected by the plant's violent passion. Justis's quirky libidinous machine seems to suggest that even our most desperate desires may be ultimately governed by forces beyond our control. In another recent work, Justis joins the opposing forces of nature and technology to produce a machine equivalent for the secret of life itself. *Chromosome* (16) recreates the helix structure of DNA. Animated with motion and bombarded with light, the abstract shapes of heredity suddenly come to life. Lacking willful self-determination, the machine seems to be a fitting metaphor for the inexorable fact of one's biological "programming." The mechanical *Chromosome* might also be a punning commentary on the effects of genetic engineering.

Like Oppenheim, Jonathan Borofsky uses mechanized sculpture to manifest the contents of the mind. His series of "chattering men" stars a larger-than-life-size wood and aluminum figure with an articulated jaw and an accompanying sound box that repeatedly utters the word "chatter." This relentless muttering, punctuated every now and then by the fragment of a Gregorian chant, is Borofsky's metaphor for his brain's continual rehashing of angers, fears, anxieties, self-doubts, and fantasies. In one piece from the series, *Chattering Man with Eight Framed Drawings* (7), Borofsky's muttering alter ego confronts a wall covered with several of his own sketches and doodles. In addition to representing the clamor of Borofsky's internal dialogues, this chattering automaton calls to mind the rhetoric of today's art discourse. It is almost as if Borofsky is satirizing the endless critical chatter which now surrounds his own work.

In the following quote, Victor Frankenstein describes the invention of a very different kind of machine:

> *With an anxiety that almost amounted to agony, I collected the instruments of life around me, that I might infuse a spark of being into the lifeless thing that lay at my feet ... I saw the dull yellow eye of the creature open; it breathed hard, and a convulsive motion agitated its limbs.*

Like Mary Shelley's impassioned Dr. Frankenstein, many artists use mechanical devices to create the illusion of life. Gerald Heffernon's bizarre rubber creatures squirm, quiver, wiggle, and twist. The artist animates his exotic organisms with light-powered photoelectric cells cleverly disguised within the creature's anatomy. Heffernon explains: "I began to see arcane connections between electromechanics and the electrochemical functioning of real living things." In *Retractor Retractor* (12), a strange pulsating sea monster hangs captive within a glass case like a specimen in a laboratory. *The Unborn Idea* (13) is a translucent white embryo that contains the squirming mass of some fantastic species. Describing the inspiration for his work, Heffernon recounts: "I grew up near a major zoo and went there often to gawk and throw peanuts. Many of the stranger creatures seemed to be inventions, like the monsters of Hieronymous Bosch. Soon I began to draw and sculpt such inventions of my own."

James Seawright's more abstract robotic "house plants" also mimic nature. Like their real-life counterparts, his mechanical plants respond to variations in light with a flurry of movement (26). In one piece, a viewer-operated light sensor triggers the twisting motion of a mysterious black orb, while across its surface "pollen" dots flicker bright yellow in a computer-generated pattern of nervous excitement. A second plant, a simple box "flower," slowly opens its square black petals to reveal grids of small red lights blinking in swiftly moving patterns. The complex, sequenced movements of Seawright's electronic blossom and its male companion possess an eerie feeling of intelligence.

The creation of machines that simulate the actions and appearance of living things has a long history. Among the earliest known devices are the articulated earthenware statuettes of ancient Egypt and the mechanical contrivances built by Hero of Alexandria around the first century A.D. Perhaps the most extraordinary examples emerge from the eighteenth century when there developed a marked taste, especially among the princely and noble circles of Europe, for marionettes, mechanical toys, animated tableaux, and moving clock figures. During that period, the mechanical genius Jacquet-Droz built his well-known *Draughtsman,* a life-size android that demonstrated the amazing ability to write and draw. The inventor Jacques de Vaucanson also created a number of highly realistic automatons, including the *Flutist, Drummer,* and *Tambourine Player.*

The contemporary singing robots of Clair Colquitt, with their uncanny lifelike movements, possess the same quality of magic as their mechanical predecessors. Colquitt's recent quartet of computer-animated performers called *Misty and the Pin Boys* (9) stars a voluptuous seven-and-one-half-foot chanteuse with dark curly hair, bulging eyes, long lashes, thick lips, and a voice reminiscent of Eartha Kitt. For just one quarter, Misty, wearing a polka-dotted disco dress, will perform one of several musical numbers in her inimitable "talking blues" style. The lead singer's repertoire of songs includes "Money Machine," "Love-a-Lotto," and "Misty's Advice," written by the artist himself, as well as the classic anthem "America the Beautiful." Accompanying the stainless steel Misty on vocals and synthesizer are the plastic Pin Boys in their plaid polyester suits and ties. (Jojo is on keyboard, while Jake and Johnny do the "shoo-ops.") With some humor, Colquitt compares his animated female figure to the mythic character Galatea, an ivory sculpture by the artist Pygmalion, who was transformed into a living woman.

The two collaborators Kristin Jones and Andrew Ginzel use mechanization to recreate the movements of the physical universe. Within darkened room-size environments, they construct miniature solar systems or surreal planetary landscapes. Their abstract scenes depict a world caught in constant flux between the primordial players — matter and energy. Hidden behind the scenes, an assortment of motors, timers, pumps, lights, fans, and projectors work to produce magnificent illusions — floating discs and rings, spinning spheres, swirling dust clouds, erupting fires, steaming volcanoes, shimmering waterfalls, diaphanous mists, glowing colors, and flickering lights (Figure 3). With such dramatic spectacles, Jones and Ginzel capture a feeling for the mysterious unseen forces which shape and govern the natural world.

In their single mechanized sculpture *Apastron* (15), Jones and Ginzel focus on the relationship between the physical forces of motion and time. Side by side within a glass reliquary, two orbs, one shimmering gold and the other ash black, spin perpetually toward one another. Beneath, a red pendulum marks and counts time, transforming the endless circular movements of the "planets" above into a series of measured moments. A stratified layering of varicolored ash behind the two revolving spheres offers a further symbol of passing time and its inexorable effects on the natural world.

Other artists employ mechanization to extremely humorous ends. In his *Thesis Machine* (6), Bryant Baker offers a funny comment on the rigors of earning an academic degree. When a viewer inserts twenty-five cents into his coffin-shaped wall sculpture, lights blink and a tiny dancing skeleton suddenly comes to life. Soon afterwards, another motor starts up and advances one more paragraph in a rolling printout of the artist's thesis paper from the University of Georgia. The entire paper can be read for about fifteen dollars. Baker's animated skeleton, jumping violently back and forth against its will, is a frequent character in his mechanized art. In *Thesis Machine,* it symbolizes the artist's ludicrous attempts to appease his teachers in graduate school.

Like Baker, Roger Machin demonstrates a macabre sense of humor. His abstract sculpture of bent and twisted steel, titled *Gwyther's Best Arm Story* (19), is an enigmatic creature with skinny attenuated limbs, a long craning neck, and a steam-shovel head. The artist drew inspiration for this absurd anthropomorphic machine from his friend Gwyther's grizzly but comic tale of a man whose arm was severed by a bulldozer. When nudged by a viewer, the metal beast wiggles as if alive. Its ungainly appendages serve as an allegory for those stories that assume strange new proportions with each retelling.

Television is the subject of humorous works by both Jim Jenkins and Rodney Alan Greenblat. In Jenkins's spooky contraption *Coordinated Programming* (14), a fleshy pink television set with mannequin arms and legs rubs its "stomach" and pats its "head." This half-human hybrid satirizes the increasingly familiar role that TV plays in contemporary life. In Greenblat's brightly painted wall sculpture *I Saw the Channel 5 in Gold* (11), a young boy's mind is possessed by "video demons." Rendered in the artist's playful cartoon style, this mechanized tableau on the diabolic effects of TV takes place within a giant television screen. Four red buttons on the control panel activate the comic book scene: eerie lights begin to glow, a house whirls around, a red devil bobs its head, and a small satellite spins in the sky amidst lavender clouds and floating TV spectres. In the center of Greenblat's candy-colored apocalypse appears a young boy whose head is filled with the golden image of Channel 5 rocking back and forth in a relentless machine movement.

Dave Quick's mechanized assemblages with their funny gadgets, neon lights, and plastic pigs combine a wry sense of humor with a concern for social issues. In his mock technological disaster *Little Nuke* (24), a top-secret rubber chicken has been plucked, stuffed with electronic gear, and suspended from the ceiling of a miniature testing lab. All of a sudden, red lights flash, bells sound, gauges spin, and the mechanical chicken releases a glowing atomic egg labeled "Little Nuke." In *Goodyear Rolling Pin* (23), a doll-size kitchen becomes the setting for a domestic nightmare parodying the almost supernatural power of today's "electronic advertising." With the push of a button, lights flash, doors open, and a number of familiar household products assume a life of their own. Suddenly, the Pillsbury Doughboy starts to glow. Then a giant hand emerges to poke the plump figure, sending it into a fit of insane laughter. Quick is a madcap tinkerer who uses modern technology as a vehicle for its own negation.

Robert Mark Packer also creates humorous machines to satirize modern technology. Many of his recent works use mechanized movement as an analogue for the dehumanizing forces of mass production. In *Hari Kari Ketchup King* (22), fifty-three plastic tomatoes, arranged in tiers atop an ornate oriental prayer shrine, commit mass hari-kari by repeatedly slicing themselves with motorized knives. Meanwhile, a miniature public-address system broadcasts the inspirational sounds of Chinese folk tunes set to disco music. Packer likens modern technology to political or religious ideologies which demand self-sacrifice and includes two Chairman Mao tomato portraits as well as a Japanese national flag with a rising tomato sun. In *The Fall of the House of Udder* (21), Packer again addresses the merciless forces of machine production. A freezer transformed into a black-and-white Holstein cow opens up to reveal twenty-four vinyl steaks which squeak and gesture pathetically. Packer's talking tenderloins are a truly off-beat protest against technology's transformation of living creatures into consumable products.

In his ironic painting *Ignorance Breeds Extinction* (10), Gregory Green uses mechanization to attack the mythology of political power. Combining an upside-down image of a dinosaur, a waving American flag, and the sounds of macabre organ music, the artist portends demise for a society of idle patriots unwilling to question and challenge the established order. The repetitive motorized flag waving, a reference to the spectacle of democracy, also symbolizes mindless obeisance to political ideology. According to Green, the traditional gold picture frame which surrounds his painting signals further compliance to convention, while the overtly displayed electrical cord suggests the presence of an invisible but controlling (societal) force. Like many artists, Green views mechanization not only as a way to extend the meaning of his art but also as a strategy to attract the viewer acclimated to the pace and patterns of contemporary society and its entertainment media.

In contrast to Green's overt political warnings, Iris Adler's small mechanized tableaux are enigmatic musings on the nature of life and death, good and evil. Delicate assemblages of roughly shaped figures, dried ferns, pressed flowers, shells, rocks, and feathers, Adler's seemingly naive depictions are powerful and menacing. In a piece titled *Sloth* (3) from the artist's series on the Seven Deadly Sins, a skeleton slowly rocks the hammock of a sleeping man. A few moments later, blinking lights and ominous buzzing sounds signal the actions of another death figure as he begins to dig the man's grave. The character of Death also plays a role in Adler's wonderfully twisted narratives from the Book of Genesis. In *Garden of Eden No. 8* (2), the artist envisions the Tree of Knowledge as a gnarled skeleton because any attempt to pluck the tree's fruit doomed one to mortality. In addition to Adam and Eve, her revisionist tale includes a serpent with flashing red eyes, two Darwinian apes, a few goats, and a bird. The actors in Adler's stories bob, squirm, and rotate in a programmed series of simple actions accompanied by flickering lights, tinkling music, and an assortment of strange electronic warbles. Reflecting on her use of hidden mechanisms and on the religious connotations in her work, the artist once mused: "Many assume that the movements in my pieces are controlled by 'God,' who is known to work in such mysterious ways."

Surrounded by machines, from electric can openers to airplanes, we forget that there is mystery and magic in their movement. The artist's eccentric machines beguile and amuse us once again.

> *No sooner had the puppet discovered that he had feet than he jumped down from the table on which he was lying and began to spring and cut a thousand capers about the room, as if he had gone mad with the greatness of his delight.*
>
> *Pinocchio* by Caro Collodi

Joanne Cubbs

Figure 2. Installation view of ECCENTRIC MACHINES exhibition.

22 Robert Mark Packer, *Hari Kari Ketchup King,* 108 x 72 x 48."

5 Alice Aycock, *Fata Morgana (Of Things Seen in the Sky),* 149 x 57 x 130"

18 Dean Langworthy, *Allegory Engine Series No. 5,* 94 x 38 x 79″

2 Iris Adler, *Garden of Eden No. 8,* 19½ x 17½ x 12."

24 Dave Quick, *Little Nuke,* 12½ x 26 x 17"

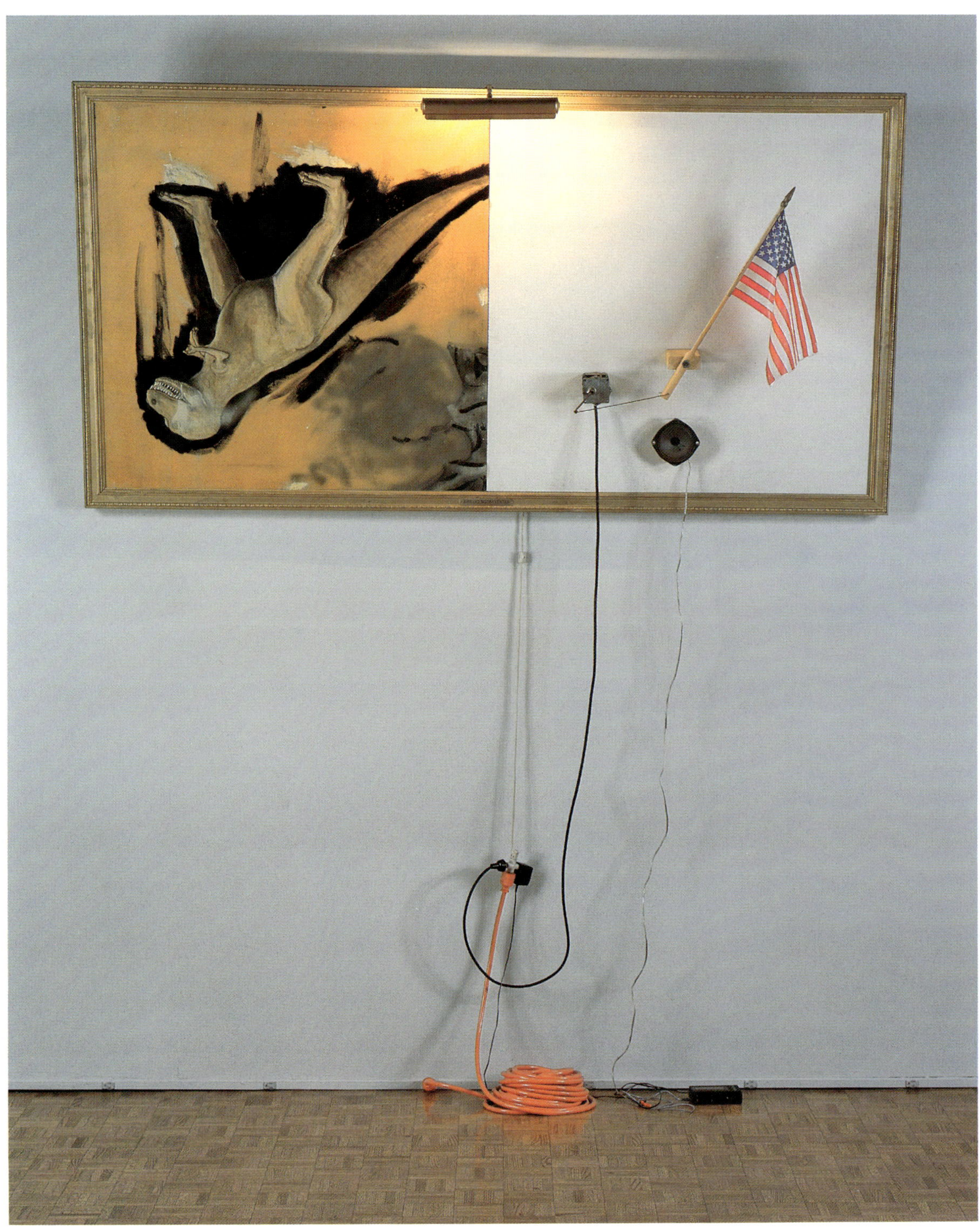

10 Gregory Green, *Ignorance Breeds Extinction,* 110 x 88½ x 18″ (installation).

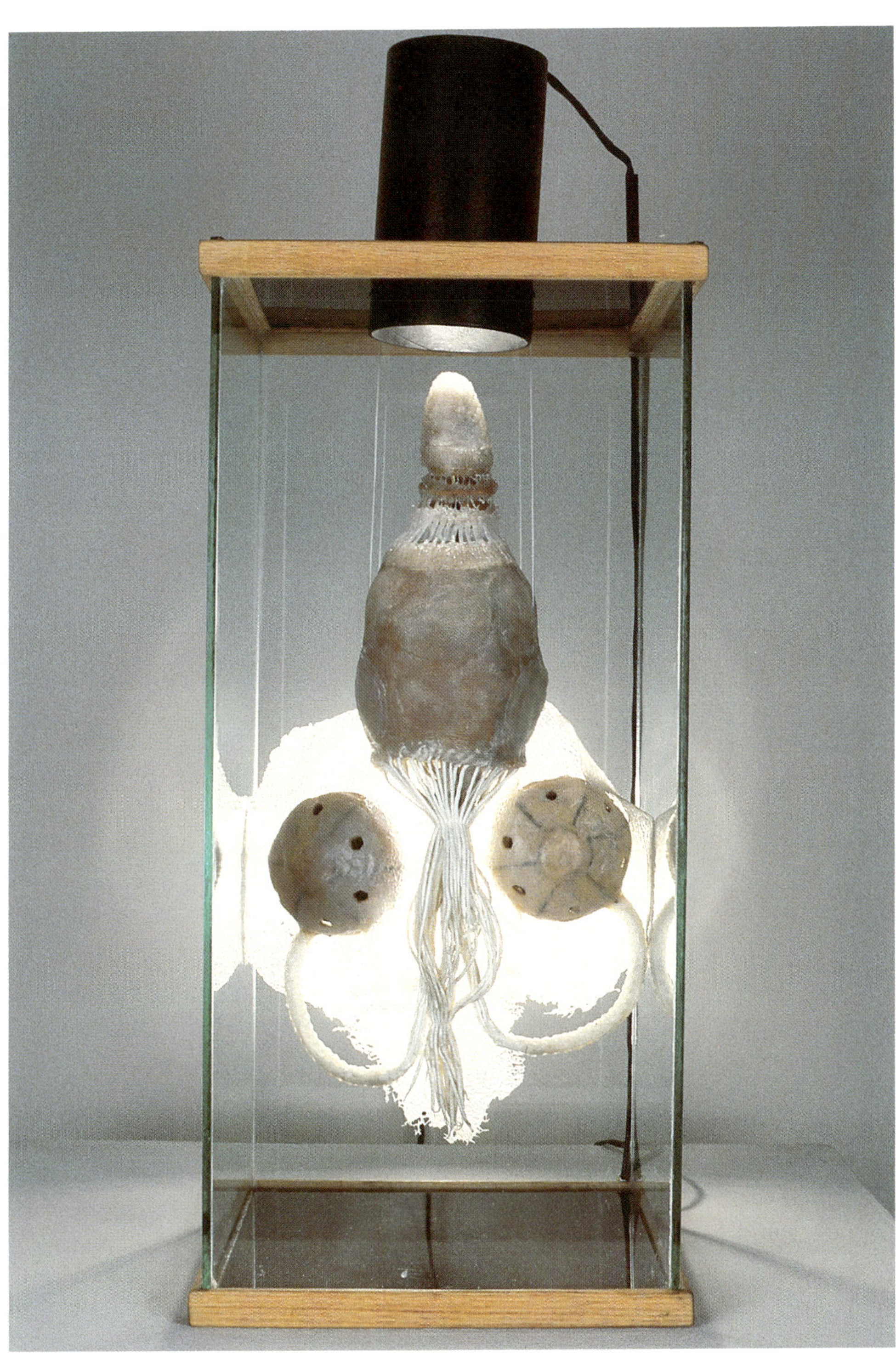

12 Gerald Heffernon, *Retractor Retractor*, 28 x 10½ x 10½"

14 Jim Jenkins, *Coordinated Programming*, 72 x 48 x 36"

26 James Seawright, *House Plants,* 22¼ x 60 x 60″ (installation).

9 Clair Colquitt, *Misty and the Pin Boys,* 90 x 144 x 144'' (installation).

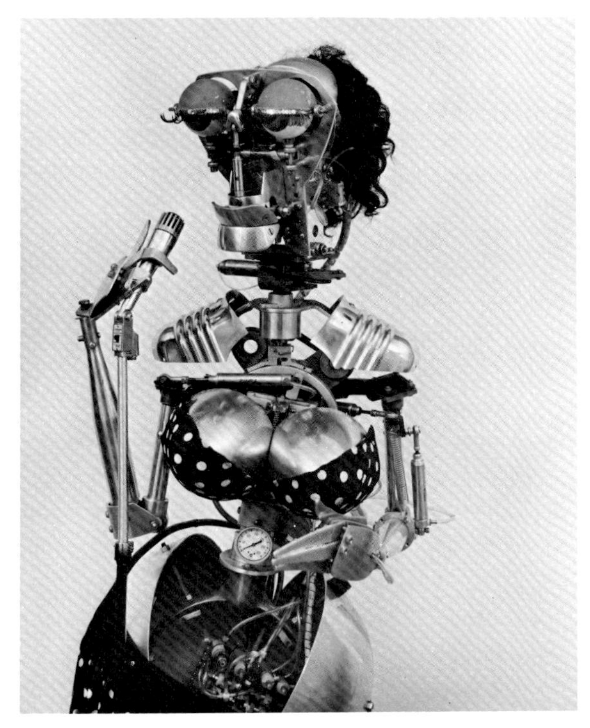
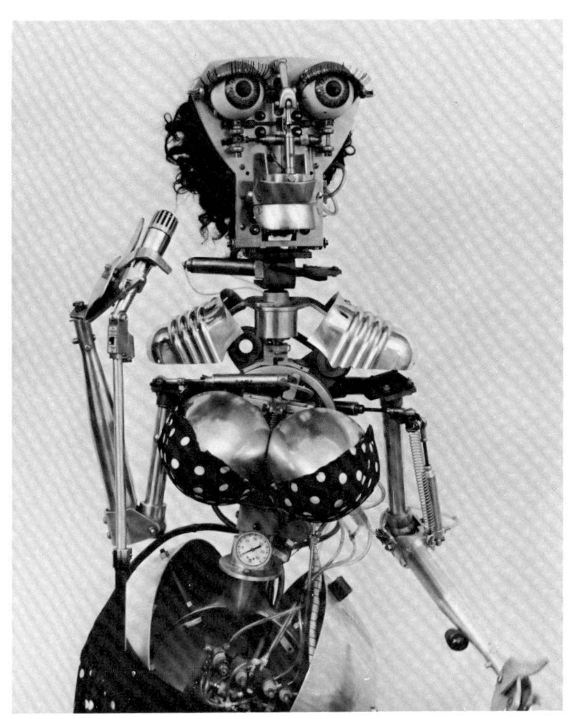

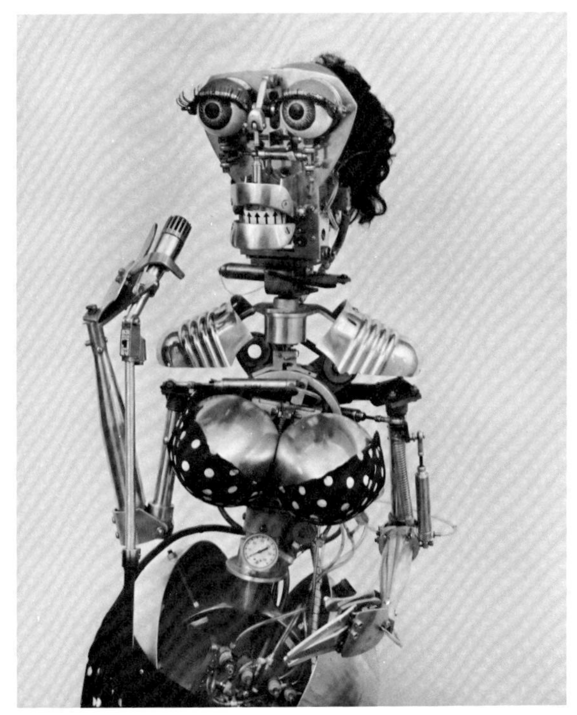

9 Details of *Misty*.

17 Gary Justis, *Controlled Stamen,* 82 x 28 x 33"

Figure 3. Kristen Jones and Andrew Ginzel, *Clepsydra*, 1986; water, coal, brass, nylon, pigment, copper, gold, motor, and electronics; 108 x 96 x 79". Installation: Lawrence Oliver Gallery, Philadelphia, Pennsylvania.

25 Dave Quick, *Pig Descending a Staircase,* 29 x 22 x 19."

21 Robert Mark Packer, *The Fall of the House of Udder,* 60 x 72 x 36."

20 Dennis Oppenheim, *The Day Before Starry Night (For Vincent Van Gogh)*, 76 x 49 x 36½"

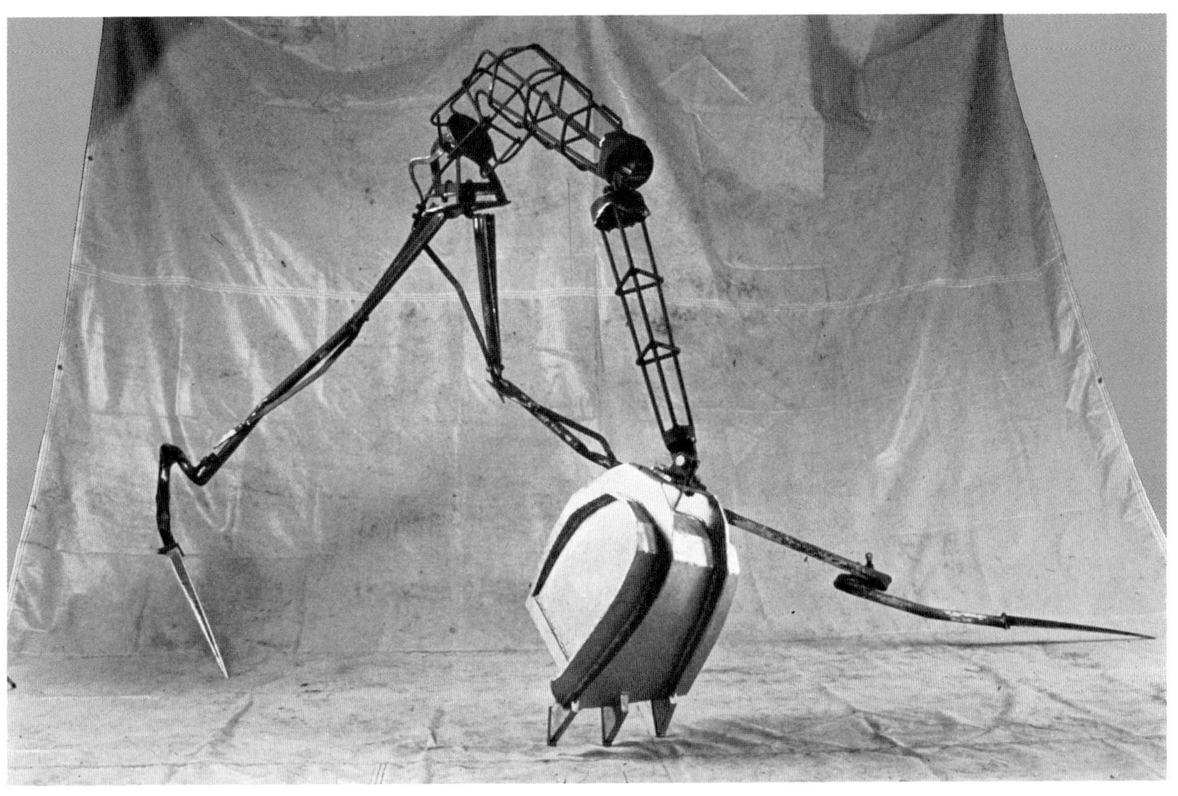

19 Roger Machin, *Gwyther's Best Arm Story,* 84 x 84 x 156."

3 Iris Adler, *Sloth,* 13 x 24 x 12″

6 Bryant Baker, *Thesis Machine,* 43 x 15½ x 5¼"

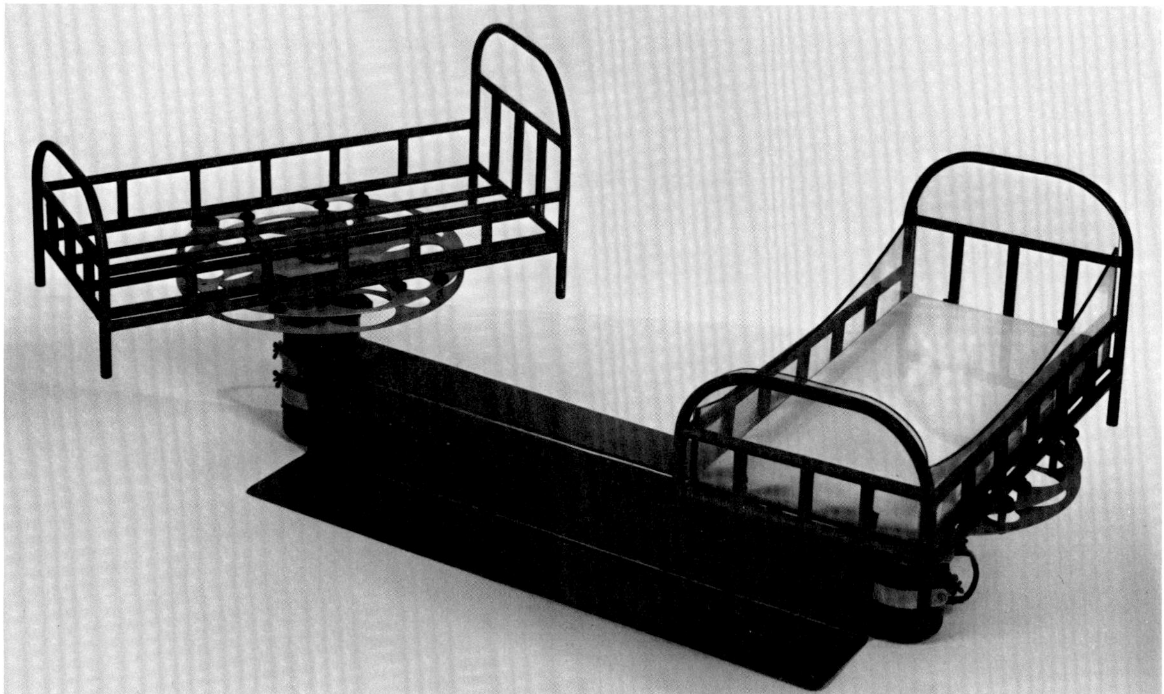

4 Lewis Alquist, *States of Matter Undergoing Habitual Rotation* (moving and static), 24 x 30 x 72."

8 Joe Cavalier, *Moon/Lightning,* 72'' height with variable diameter of 24 to 60''.

7 Jonathan Borofsky, *Chattering Man with Eight Framed Drawings*, 82½ x 24 x 13″ (figure).

CATALOGUE

Dimensions are in inches followed in parentheses by centimeters. Height precedes width precedes depth. Works have been loaned by the artist unless otherwise noted.

Iris Adler
1 GARDEN OF EDEN NO. 3
1984
Mixed media including found natural objects, lights, sound, electric motors, electronics, 13 x 15 x 12" (33.0 x 38.1 x 30.5 cm)

Iris Adler
2 GARDEN OF EDEN NO. 8
1984
Mixed media including found natural objects, lights, sound, electric motors, electronics, 19½ x 17½ x 12" (49.5 x 44.5 x 30.5 cm)

Iris Adler
3 SLOTH
1986
Mixed media including found natural objects, lights, sound, electric motors, electronics, 13 x 24 x 12" (33.0 x 61.0 x 30.5 cm)

Lewis Alquist
4 STATES OF MATTER UNDERGOING HABITUAL ROTATION
1983
Welded steel bed frames, glass, electric motors, milk substitute, 24 x 30 x 72" (61.0 x 76.2 x 182.9 cm)

Alice Aycock
5 FATA MORGANA (OF THINGS SEEN IN THE SKY)
1984
Steel, glass, mirror, neon light, grinding wheel, electric motors, 149 x 57 x 130" (378.5 x 144.8 x 330.2 cm)
Courtesy of John Weber Gallery, New York

Bryant Baker
6 THESIS MACHINE
1986
Found objects (cast, brazed, and assembled), light, sound, electromagnets, electric motors, 43 x 15½ x 5¼" (109.2 x 39.4 x 13.3 cm)

Jonathan Borofsky
7 CHATTERING MAN WITH EIGHT FRAMED DRAWINGS
1986
Aluminum, wood, primer, bonds, electric motor, speakers; ink, colored pencil, pencil, and marker on paper; Chattering Man 82½ x 24 x 13" (209.6 x 61.0 x 33.0 cm); framed drawings 9½ x 6" (24.1 x 15.2 cm) to 12 x 9" (30.5 x 22.9 cm)
From a private collection

Joe Cavalier
8 MOON/LIGHTNING
1985
Painted steel, lucite, wire cable, electric motor, 72 x 24" to 72 x 60" (182.9 x 61.0 to 182.9 x 152.4 cm)

Clair Colquitt
9 MISTY AND THE PIN BOYS
1986
Aluminum, steel, neoprene, electric motors, computer-controlled mechanisms; Misty 91¾ x 28½ x 36¾" (233.0 x 72.4 x 93.3 cm); Jojo with keyboard 58½ x 31¼ x 31½" (148.6 x 79.4 x 78.7 cm); Jake and Johnny 56¼ x 36¼ x 18" (142.9 x 92.1 x 45.7 cm); installation 90 x 144 x 144" (228.6 x 365.8 x 365.8 cm). For one quarter, Misty performs one of the following musical numbers: "Do the Misty" by Gyda Fossland, "Money Machine" by Clair Colquitt, "Love-a-lotto" by Clair Colquitt, "Misty's Advice" by Clair Colquitt, "Rickey's Song" by Gyda Fossland, or "America the Beautiful."

Gregory Green
10 IGNORANCE BREEDS EXTINCTION
1984
Mixed media, found painting, picture light, electric motor, extension cord, tape recorder, speaker; painting 46 x 88½" (116.8 x 224.8 cm); installation 110 x 88½ x 18" (279.4 x 224.8 x 45.7 cm)

Rodney Alan Greenblat
11 I SAW THE CHANNEL 5 IN GOLD
1987
Acrylic on wood, mixed media, lights, electric motor, 66 x 66 x 7" (167.6 x 167.6 x 17.8 cm)
From the collection of Martin Edelston, New York; courtesy of Gracie Mansion Gallery, New York

Gerald Heffernon
12 RETRACTOR RETRACTOR
1985
Silicone rubber, photovoltaic cells, electric motor, glass and oak display case, brass lamp, 28 x 10½ x 10½" (71.1 x 26.7 x 26.7 cm)

Gerald Heffernon
13 THE UNBORN IDEA
1985
Silicone rubber, photovoltaic cell, electric motor, plexiglass and wood display case, reflector beads, porcelain lamp, 14¾ x 18½ x 12" (37.5 x 47.0 x 30.5 cm)

Jim Jenkins
14 COORDINATED PROGRAMMING
1987
Television set, mannequin arms and legs, electric motor, 72 x 48 x 36" (182.9 x 121.9 x 91.4 cm)

Kristin Jones and **Andrew Ginzel**
15 APASTRON
1986
Glass, steel, ash, cinders, gold, paint, electric motors; sculpture 73¼ x 44½ x 24½" (186.1 x 113.0 x 62.2 cm); background 108 x 108" (274.3 x 274.3 cm)

Gary Justis
16 CHROMOSOME
1986
Aluminum, plastic, lights, timers, electric motors, 98 x 40 x 40" (248.9 x 101.6 x 101.6 cm)

Gary Justis
17 CONTROLLED STAMEN
1986
Aluminum, plastic, lights, timers, electric motors, 82 x 28 x 33" (208.3 x 71.1 x 83.8 cm)

Dean Langworthy
18 ALLEGORY ENGINE SERIES NO. 5
1984
Steel rails and chain, limestone block, wooden stool, crank, gearing, 94 x 38 x 79" (238.8 x 96.5 x 200.7 cm)
Courtesy of Paul Klein Gallery, Illinois

Roger Machin
19 GWYTHER'S BEST ARM STORY
1986
Fabricated steel, galvanized steel, brass, 84 x 84 x 156" (213.4 x 213.4 x 396.2 cm)

Dennis Oppenheim
20 THE DAY BEFORE STARRY NIGHT
(FOR VINCENT VAN GOGH)
1983
Wood, fiberglass, expanded steel, galvanized steel, enamel, lights, electric motor, 76 x 49 x 36½" (193.0 x 124.5 x 92.7 cm)
Courtesy of Sander Gallery, New York

Robert Mark Packer
21 THE FALL OF THE HOUSE OF UDDER
1984
Mixed media, painted freezer, plastic chew toys, lights, sound, electric motor, 60 x 72 x 36" (152.4 x 182.9 x 91.4 cm)

Robert Mark Packer
22 HARI KARI KETCHUP KING
1984
Mixed media, plastic tomatoes, photographs, flags, lights, sound, electric motors, 108 x 72 x 48" (274.3 x 182.9 x 121.9 cm)

Dave Quick
23 GOODYEAR ROLLING PIN
1982
Mixed media including found objects, illuminated photographs, lights, sound, timers, electric motors, 28 x 23 x 17½" (71.1 x 58.4 x 43.2 cm)

Dave Quick
24 LITTLE NUKE
1979
Mixed media including found objects, industrial gadgetry, illuminated photographs, lights, sound, timers, electric motors, electronics, 12½ x 26 x 17" (31.8 x 66.0 x 43.2 cm)

Dave Quick
25 PIG DESCENDING A STAIRCASE
1985
Mixed media including found objects, electric motor, 29 x 22 x 19" (73.7 x 55.9 x 48.3 cm)

James Seawright
26 HOUSE PLANTS
1983
Metal, plastic, electric motor, electronic parts, computer-controlled mechanism, light sensor; square plant 22¼ x 15 x 17" (56.5 x 38.1 x 43.2 cm); rounded plant 16½ x 10 x 12" (41.9 x 25.4 x 30.5 cm); control box 12 x 19 x 12" (30.5 x 48.3 x 30.5 cm); installation 22¼ x 60 x 60 " (56.5 x 152.4 x 152.4 cm)

Iris Adler
Highland Park, Illinois

Iris Adler received a Master of Fine Arts degree from The School of the Art Institute of Chicago in 1981. Since then, her work has been presented in group exhibitions at numerous galleries and museums, including The Art Institute of Chicago, Northeastern Illinois University, Hyde Park Art Center, and N.A.M.E. Gallery in Chicago, Illinois; the Carnegie Arts Center in Covington, Kentucky; Hallwalls in Buffalo, New York; the Evanston Art Center in Evanston, Illinois; and WARM Gallery in Minneapolis, Minnesota. The artist's mechanized assemblages were the subject of one-person exhibitions at Artemesia Gallery in Chicago in both 1984 and 1985. Adler was awarded grants from the Illinois Arts Council in 1985 and 1986.

Lewis Alquist
Tempe, Arizona

Lewis Alquist is currently associate professor of art at Arizona State University in Tempe. In 1972, he received a Master of Fine Arts degree in sculpture from the Cranbrook Academy of Art in Bloomfield Hills, Michigan. From 1973 to 1980, he was an assistant professor at Edinboro State College in Edinboro, Pennsylvania. In 1980, he served as visiting faculty member at The School of the Art Institute of Chicago. During the last seven years, the artist's work has been featured in numerous exhibitions in Illinois, Arizona, and Pennsylvania. In 1986, he completed a sculpture for permanent exhibition at The Exploratorium in San Francisco, California. Alquist was the recipient of fellowships from the Illinois Arts Council in 1984 and 1985 and from the National Endowment for the Arts in 1986.

Alice Aycock
New York, New York

Alice Aycock is a sculptor of international repute. Born in Harrisburg, Pennsylvania, the artist received a Master of Arts degree from Hunter College in New York in 1971. Since the late 1970s, her drawings, sculpture, and installation works have been featured in countless exhibitions at major galleries and museums throughout this country and abroad. In addition to frequent solo showings at John Weber Gallery in New York City, the artist's work has been repeatedly included in Italy's prestigious Venice Biennale and in the Biennial exhibition of the Whitney Museum of American Art in New York City. A 1983-84 retrospective of Aycock's work that traveled to five cities in West Germany, Holland, and Switzerland underscored further her importance as a major twentieth-century American artist.

Bryant Baker
Athens, Georgia

Bryant Baker received a Master of Fine Arts degree in sculpture from the University of Georgia in Athens in 1986. His one-person exhibitions include showings at Ohio State University in Columbus in 1982 and at the Tate Gallery of the University of Georgia in 1986. Since 1980, the artist's work has been shown in group exhibitions throughout Georgia and in ten other states. Baker, who served as president of the Georgia Sculptor's Society in 1985 and 1986, has developed a strong body of mechanized work. He has noted that, "A nonkinetic piece is like a photograph of a past event or idea, but a work in motion is current."

Jonathan Borofsky
Venice, California

Jonathan Borofsky is one of the most widely known artists in the United States today. Although involved with conceptual art investigations during the 1960s, Borofsky is widely acclaimed for his bold figurative work. A master of "gesamtkunstwerk," the artist often combines drawings, paintings, graffiti, visual and verbal notes, audiotapes, projected images, and both static and kinetic sculpture. He has created over fifty installations in major museums in the United States and abroad. Borofsky received a Master of Fine Arts degree from the Yale School of Art and Architecture in 1966 and has been represented in numerous one-person exhibitions at Paula Cooper Gallery in New York City since the late 1970s. In both 1981 and 1983, his work appeared in the prestigious Biennial exhibition at New York's Whitney Museum of American Art. In 1984, the Philadelphia Museum of Art in Pennsylvania mounted a major traveling retrospective of his work.

Joe Cavalier
Chicago, Illinois

Joe Cavalier is a professor at The School of the Art Institute of Chicago, where he also received a Master of Fine Arts degree in 1968. In 1975, he was awarded a Fulbright-Hayes Exchange Scholars Grant to teach in Brighton, England. In recent years, his sculpture has been included in numerous exhibitions throughout Illinois. The kinetic piece *Cloud Catcher* was featured in "Sculpture Chicago '83," and in 1986, the artist was represented in "Making Waves," an interactive art/science exhibition at the Evanston Art Center in Evanston, Illinois. Cavalier is also a professional jazz cornetist who has recorded and performed internationally with the Red Rose Ragtime Band.

Clair Colquitt
Seattle, Washington

Clair Colquitt received a Master of Fine Arts degree in ceramics from the University of Washington in Seattle in 1969. In 1976, he was awarded a fellowship from the National Endowment for the Arts. Since the mid-1970s, his work has been widely exhibited throughout the Seattle area. In recent years, Colquitt has also undertaken public art commissions and organized special collaborative arts projects. In 1982, he designed and coordinated "Artarcade," a coin-operated show involving the work of thirty artists at Seattle's "Bumbershoot" exposition. The exhibition included altered pinball machines, adult "kiddie" rides, and video games. In the proposal for his coin-operated exhibition, Colquitt wrote: "This modern era has proven that anyone will insert a quarter into anything ... why not art?"

Gregory Green
Chicago, Illinois

Gregory Green received a Bachelor of Fine Arts degree from the Art Academy of Cincinnati in 1981 and a Master of Fine Arts from The School of the Art Institute of Chicago in 1984. Green, a sculptor as well as a performance and installation artist, uses mechanization to integrate elements of performance into his sculpture. Since 1980, his work has been featured in group exhibitions at museums and galleries throughout the Midwest, among them the Museum of Contemporary Art in Chicago, Illinois, and the Cincinnati Museum of Art and Cincinnati Contemporary Arts Center in Ohio. Dart Gallery in Chicago developed a one-person exhibition of his work in 1985. His performance and installation pieces have appeared at C.A.G.E. Gallery in Cincinnati and at N.A.M.E. Gallery, Randolph Street Gallery, Feature, the Peace Museum, the Chicago International Art Expo, and Federal Plaza in Chicago.

Rodney Alan Greenblat
New York, New York

Rodney Alan Greenblat is a widely known artist from New York's East Village. Born in Daly City, California, he received a Bachelor of Fine Arts degree from the School of Visual Arts in New York in 1982. Shortly afterward, he became affiliated with Gracie Mansion Gallery in Manhattan. Within the last few years, his work has been featured in numerous exhibitions at museums and galleries throughout this country and in Australia, Canada, Denmark, Italy, Sweden, and West Germany. In 1985, he was selected for inclusion in the Biennial exhibition at the Whitney Museum of American Art in New York City. He has had recent solo exhibitions at Gracie Mansion Gallery; Pennsylvania State University Museum of Art in University Park; Karl Bornstein Gallery in Los Angeles, California; the Australian Centre for Contemporary Art in Victoria, Australia; and Anna Friebe Galerie in Cologne, Germany. His work has also been featured in group exhibitions at such institutions as The Museum of Modern Art, Queens Museum, and Holly Solomon Gallery in New York City; the Virginia Museum of Fine Arts in Richmond; the University Art Museum at the University of California, Santa Barbara; and the Indianapolis Museum of Art in Indiana.

Gerald Heffernon
Davis, California

Gerald Heffernon received a Bachelor of Arts degree from the University of Illinois at Urbana. Since the late 1970s, his work has been featured in exhibitions at Joseph Chowning Gallery and the San Francisco Arts Commission Urban Fair in San Francisco, California, and Paul Waggoner Gallery in Chicago, Illinois, among others. During the past year, his extraordinary creatures have also traveled to Avignon, Paris, and Caen in France as part of the exhibition "Robots-Sculptures: Les Machines Sentimentales." The artist has had permanent installations of his work at Florin Park in Sacramento, California; at Wexford Ridge in Madison, Wisconsin; and at Kiel Municipal Park in Kiel, Wisconsin. He has been the recipient of grant awards from the Sacramento Arts Commission, the Wisconsin Arts Board, and the City of Madison, Wisconsin.

Jim Jenkins
Pasadena, California

Jim Jenkins received a Master of Fine Arts degree from Syracuse University at Syracuse, New York, in 1981. For the last six years, he has served as associate professor of art at California State University at Fullerton. In 1982, the Museum of Neon Art in Los Angeles, California, sponsored a solo showing of his work and featured other pieces in group exhibitions in both 1984 and 1985. In 1983, he was given a one-person showing at Lambert-Miller Gallery in Phoenix, Arizona. Since then, Jenkins's work has been exhibited at several galleries and museums, including the Exploratorium Gallery at California State University in Los Angeles; Weber State College Gallery in Ogden, Utah; and Arizona State University at Tempe. In 1982, the artist was also commissioned to create a large-scale kinetic sculpture for the film *Crackers* directed by Louis Malle and released by Universal Films in 1984.

Kristin Jones and **Andrew Ginzel**
New York, New York

Kristin Jones, who was born in Washington, D.C., received a Bachelor of Fine Arts degree in 1979 from the Rhode Island School of Design in Providence and a Master of Fine Arts degree in 1983 from Yale University in New Haven, Connecticut. Andrew Ginzel, originally from Chicago, attended Bennington College in Bennington, Vermont, as well as The State University of New York. Jones and Ginzel have collaborated on numerous kinetic installations since 1985. Their work has comprised solo exhibitions at Art Galaxy and the New Museum of Contemporary Art in New York City; the Virginia Museum of Fine Arts in Richmond; and Lawrence Oliver Gallery in Philadelphia, Pennsylvania. They have participated in group exhibitions in New York City at the Whitney Museum of American Art at Philip Morris, Creative Time, Inc., and the Institute for Art and Urban Resources at the Clocktower. They were also recently represented in a light works exhibition at the Centre International d'Art Contemporain de Montreal in Canada. Jones and Ginzel are the 1986 recipients of a National Endowment for the Arts fellowship and a New York State Council on the Arts project award. They were selected to create a special work for *Artforum* magazine.

Gary Justis
Chicago, Illinois

Gary Justis, who was born in Maize, Kansas, received a Master of Fine Arts degree from The School of the Art Institute of Chicago in 1979. Since then, his work has been the subject of many solo exhibitions at galleries and museums throughout Chicago, including The Chicago Public Library Cultural Center in 1985, Marianne Deson Gallery in 1983 and 1986, the Museum of Science and Industry in 1982, and ARC Gallery in 1979. In 1986, he was also awarded one-person exhibitions at the Alexandria Museum of Art in Alexandria, Louisiana, and at Tower Fine Arts Center in Brockport, New York. Justis's sculptures have been featured in group exhibitions at numerous museums and galleries, among them the Whitney Museum of American Art at Philip Morris and the New Museum in New York City; the Museum of Contemporary Art, The Art Institute of Chicago, and N.A.M.E. Gallery in Chicago; and the Indianapolis Museum of Art in Indianapolis, Indiana. In 1984, Justis was awarded fellowships from the National Endowment for the Arts and the Illinois Arts Council. His work is represented in several public and private collections, including those of the Museum of Contemporary Art in Chicago; the Illinois State Museum in Springfield; and the Alexandria Museum of Art in Alexandria, Louisiana. Justis, whose father is a plumbing contractor and part-time inventor, has been building and wiring machines since he was a boy. Today his mechanized sculptures or "hyperfunctional icons" have begun to win him important critical recognition.

Dean Langworthy
Chicago, Illinois

Dean Langworthy received an undergraduate arts degree from the University of Arizona, Tucson, in 1976 and a Master of Fine Arts degree from The School of the Art Institute of Chicago in 1979. Since the late 1970s, his sculpture and installation pieces have been featured in numerous exhibitions throughout the Chicago area. For several years, his work has been included in the Chicago International Art Expo's "Mile of Sculpture." He was also commissioned to create a large-scale outdoor piece in conjunction with the "Sculpture Chicago" project in 1983. Other exhibitions include group showings at Randolph Street Gallery and Northwestern University in 1980 and at Paul Klein Gallery in 1984. Outside of Chicago, Langworthy has created major on-site pieces at Artpark in Lewiston, New York, and at the Houston Festival in Houston, Texas. His work is on permanent display at the Lakeside Foundation for the Arts in Lakeside, Michigan, and at the University of Wisconsin in Green Bay, Wisconsin. It is also represented in the collection of the Prudential Life Insurance Co. in New York City. Langworthy is a 1986 recipient of a grant from the Illinois Arts Council.

Roger Machin
Chicago, Illinois

Roger Machin, who was born in Leicester, England, received a Master of Fine Arts degree from The Art Institute of Chicago in 1980. Since 1979, he has had solo exhibitions in Chicago, Illinois, at the Rainbo Club, Marianne Deson Gallery, DePaul University, and N.A.M.E. Gallery. In 1982, he was also given a one-person exhibition at Foster Gallery at the University of Wisconsin in Eau Claire. His work has been featured in several group exhibitions, including "Sculpture Chicago '84" and the Chicago International Art Expo's "Mile of Sculpture" in 1983, 1984, and 1985. He has created installations at Artpark in Lewiston, New York, and at the Limelight Club and Randolph Street Gallery in Chicago. Other recent group exhibitions include showings at the Rockford Art Museum, Rockford, Illinois, and The Gallery in Brighton, England. Machin has received several awards, including a fellowship from the National Endowment for the Arts in 1986 and grants from the Illinois Arts Council in 1984, 1985, and 1986.

Dennis Oppenheim
New York, New York

Dennis Oppenheim has been an important figure in contemporary art since the 1960s. He attended the California College of Arts and Crafts in Oakland as an undergraduate and then, in 1965, received a Master of Fine Arts degree from Stanford University in Palo Alto, California. Shortly afterwards, he moved to New York City where he has lived ever since. During the late 1960s, Oppenheim won critical acclaim for his major contributions to the Earthworks, Body Art, and Conceptual Art movements. For the last fifteen years, his work has been featured in solo and group showings at some of the most prestigious museums throughout the United States and abroad. The artist has participated in Italy's renowned Venice Biennale and has been awarded repeated inclusion in the Biennial exhibition of the Whitney Museum of American Art in New York City. Oppenheim and his work are the subject of numerous books, catalogues, and critical reviews.

Robert Mark Packer
Grand Rapids, Michigan

Robert Mark Packer is currently an artist-in-residence at Race Street Gallery in Grand Rapids, Michigan. He received his Master of Fine Arts degree from The School of the Art Institute of Chicago in 1982. In 1983, he was awarded a one-year residency at the Roswell Museum and Art Center in Roswell, New Mexico, where he produced a major body of mechanized art. During that same year, he was the recipient of a grant from the Illinois Arts Council. In 1984, he received a grant from the Michigan Council for the Arts. During 1986, Packer served as an instructor at Hope College in Holland, Michigan, and as a visiting artist at Weber State College in Ogden, Utah. Since the late 1970s, his work has been included in group and one-person exhibitions at museums and galleries throughout the Midwest, including The Art Institute of Chicago and N.A.M.E. Gallery in Chicago, Illinois; SPACES Gallery in Cleveland, Ohio; the Pontiac Arts Center in Pontiac, Michigan; and Detroit Artists Market in Detroit, Michigan. In 1984, the artist was also represented in exhibitions at the Visual Arts Center of Alaska in Anchorage and at the Roswell Museum and Art Center in conjunction with his residency there.

Dave Quick
Los Angeles, California

Dave Quick, who was born in Springfield, Ohio, has been a resident of Los Angeles since 1968. He received his Bachelor of Arts degree in 1972 from the University of California, Los Angeles, where he has also served as a lecturer. Quick's mechanized assemblages have been shown at galleries and museums throughout California, including 1986 solo showings at the Triton Museum of Art in Santa Clara and at the Thinking Eye Gallery in Los Angeles. His work has also been included in group exhibitions at the Museum of Neon Art in Los Angeles, the Los Angeles Institute of Contemporary Art, and the Exploratorium Gallery of California State University, Los Angeles. In 1986, the artist was featured in the traveling exhibition "Southern California Assemblage" organized by the Santa Barbara Contemporary Arts Forum in Santa Barbara, California. In that same year, his work was a part of the exhibition "Neon Art" organized by Seibu Galleries for travel to the cities of Tokyo and Chiba in Japan.

James Seawright
New York, New York

James Seawright has earned much critical recognition since the late 1960s for his inventive combinations of art and technology. Born in Jackson, Mississippi, the artist is currently the director of visual arts at Princeton University in Princeton, New Jersey. Seawright, who has long been interested in the possibilities of motion and change in sculpture, began to explore the use of computers to control such effects around 1970. According to him: "The desire to animate the physical world is as old as civilization — or art. It is only in the last two or three decades that the degree of control available to artists has approached or exceeded the artist's ability to conceive of ideas." Seawright's work has been exhibited throughout the United States and abroad and is represented in the permanent collections of The Museum of Modern Art, the Whitney Museum of American Art, and the Guggenheim Museum in New York City as well as many other museums throughout the country.

Typeset by Schwartz Typesetting Co., Sheboygan, Wisconsin
Typestyle: Avant Garde
Printed by Universal Lithographers, Inc., Sheboygan, Wisconsin
Cover: 85# Reflections gloss enamel cover
Text: 90# Reflections gloss enamel book

Photo credits:

Jeff Atherton, catalogue number 24.
Jon Bolton, catalogue numbers 2, 3, 4, 6, 8, 9, 10, 11, 14, 21, 22, 25, and 26.
Eric Dean, catalogue number 7.
Matt Dinerstein, catalogue number 19.
T. Charles Erickson, figure 3.
Eric Johnson, catalogue numbers 12 and 18, figures 1 and 2.
Sander Gallery, catalogue number 20.
Tom Van Eynde, catalogue numbers 16 and 17.
John Weber Gallery, catalogue number 5.